INCREDIBLE ANIMAL FACE-OFFS

ANIMAL BUILDER BRAWL

ELSIE OLSON

Consulting Editor, Diane Craig, M.A./Reading Specialist

Super Sandcastle

An Imprint of Abdo Publishing
abdobooks.com

abdobooks.com

Published by Abdo Publishing, a division of ABDO, PO Box 398166, Minneapolis, Minnesota 55439.

Printed in the United States of America, North Mankato, Minnesota
102019
012020

THIS BOOK CONTAINS RECYCLED MATERIALS

Design: Sarah DeYoung, Mighty Media, Inc.
Production: Mighty Media, Inc.
Editor: Jessica Rusick
Cover Photographs: Shutterstock Images
Interior Photographs: Shutterstock Images, pp. 4, 5, 6, 7, 8, 9, 10, 11, 13 (bottom right), 14, 15, 16 (inset), 17, 18, 19, 20, 21, 22, 23 (inset); Universal Images Group via Getty Images, p. 7 (bottom right); wilpunt/iStockphoto, p. 23; Yoji Okata/Nature Production/ Minden Pictures, p. 5 (top right), 12, 13, 16

Library of Congress Control Number: 2019943204

Publisher's Cataloging-in-Publication Data
Names: Olson, Elsie, author.
Title: Animal builder brawl / by Elsie Olson
Description: Minneapolis, Minnesota : Abdo Publishing, 2020 | Series: Incredible animal face-offs
Identifiers: ISBN 9781532191923 (lib. bdg.) | ISBN 9781532178726 (ebook)
Subjects: LCSH: Animals--Habitations--Juvenile literature. | Wild animal dwellings--Juvenile literature. | Animals, Habits and behavior of--Juvenile literature. | Social behavior in animals--Juvenile literature.
Classification: DDC 591.564--dc23

Super Sandcastle™ books are created by a team of professional educators, reading specialists, and content developers around five essential components—phonemic awareness, phonics, vocabulary, text comprehension, and fluency—to assist young readers as they develop reading skills and strategies and increase their general knowledge. All books are written, reviewed, and leveled for guided reading, early reading intervention, and Accelerated Reader™ programs for use in shared, guided, and independent reading and writing activities to support a balanced approach to literacy instruction.

CONTENTS

BATTLE OF THE BUILDERS

HONEYBEE

The animal kingdom is full of stars. But some animals stand out. These animals are the best builders.

Their creations provide shelter. They **attract** mates. And some help animals raise their young.

Animal engineers are all around us. But what if you matched them up in face-offs? Which animal would be the best builder?

BOWERBIRD

TERMITE
WHITE-SPOTTED PUFFERFISH
BEAVER
SOCIABLE WEAVER

CONSTRUCTION CONFLICT

Beavers are skilled builders. Termites are too! Which animal would win a construction conflict?

BEAVER

DURABLE DAMS

This **rodent's** dams can be seen from space. Meet the bulky beaver!

BEAVER STATS

HOME
Forests near fresh water

FOOD
Leaves, bark, and water plants

SIZE
About 60 pounds (27 kg) and up to 51 inches (130 cm) long

A BEAVER IS BIGGER THAN A HOUSE CAT.

BEAVER LODGE

TERMITE

MAGNIFICENT MOUNDS

This insect's mounds tower over the **savanna**. Say hello to the terrifying termite!

TERMITE STATS

HOME
Africa, Australia, South America, and Asia

FOOD
Wood, plant material, and **fungus**

SIZE
Less than ½ inch (1 cm) long

AN AVERAGE TERMITE IS SMALLER THAN A DIME.

TERMITE DIME

CONSTRUCTION CONFLICT

BEAVER VS TERMITE

BEAVER

Beavers build dams to form ponds. They build lodges in the ponds. Each lodge is home to a beaver family.

HOW IT'S MADE

Dams are made of logs, branches, and mud. Lodges are also made of branches and mud.

THAT'S BIG!

One beaver dam in Canada is 2,790 feet (850 m) long. It can be seen from space!

TERMITE

Each mound can house millions of termites. These critters live in nests beneath the mound.

HOW IT'S MADE

Mounds are made of soil, **dung**, and **saliva**. They can take years to build.

THAT'S BIG!

Termite mounds can be up to 30 feet (9 m) high.

DECORATOR DUEL

White-spotted pufferfish are dazzling decorators. Bowerbirds are too! Which animal would win a decorator duel?

WHITE-SPOTTED PUFFERFISH

STYLISH CIRCLE SHAPER

This fish's nests look like underwater circles. Meet the artistic white-spotted pufferfish!

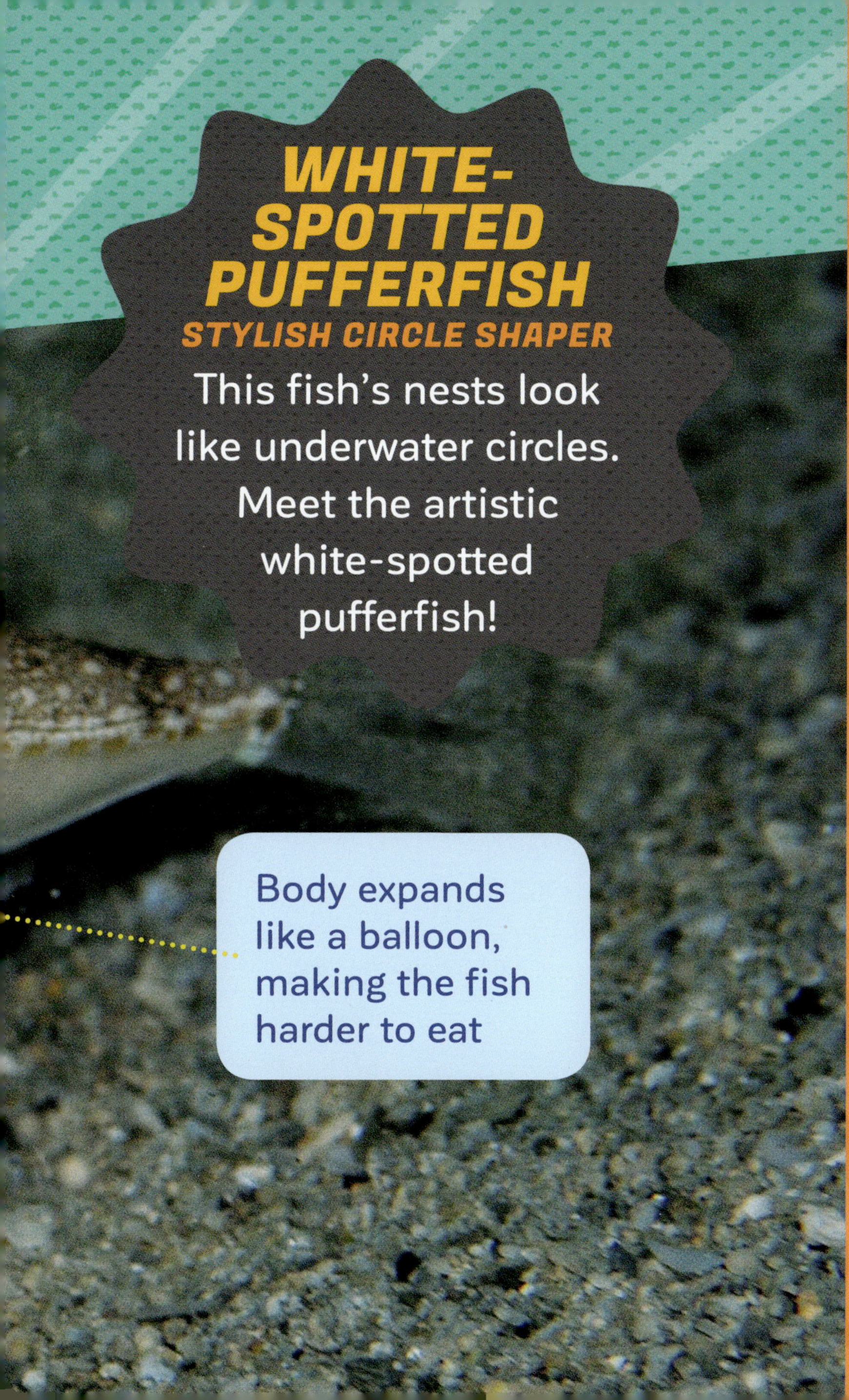

WHITE-SPOTTED PUFFERFISH STATS

HOME
Oceans near Japan

FOOD
Small marine animals and **algae**

SIZE
About 5 inches (13 cm) long

A WHITE-SPOTTED PUFFERFISH IS ABOUT AS LONG AS A SODA CAN.

BOWERBIRD

MARVELOUS MAKER

This bird decorates structures to **attract** a mate. Get on your feet for the brilliant bowerbird!

Male bowerbirds with dull feathers build fancier **bowers** than males with bright feathers

BOWERBIRD STATS

HOME

Australia and the South Pacific island of New Guinea

FOOD

Fruit, seeds, and insects

SIZE

Between 9 and 15 inches (23 and 38 cm) long

A BOWERBIRD IS SMALLER THAN A HOUSE CAT.

DECORATOR DUEL

WHITE-SPOTTED PUFFERFISH VS BOWERBIRD

WHITE-SPOTTED PUFFERFISH

White-spotted pufferfish nests are circular patterns in the sand. Males build the nests to **attract** females.

HOW IT'S MADE

Male fish flap their fins on the ocean bottom. The movement creates sand patterns.

THAT'S BEAUTIFUL!

White-spotted pufferfish decorate their circles with shells and coral.

BOWERBIRD

Male bowerbirds build **bowers** on the forest floor. A female visits many bowers. Then she chooses a mate.

HOW IT'S MADE

Bowerbirds weave bowers from twigs and sticks.

THAT'S BEAUTIFUL!

Males decorate their bowers with shells and flowers. They use human items too. This includes bottle caps and car keys!

COLONY CONTEST

Honeybees are unbelievable builders. Sociable weaver birds are too! But who would win a colony contest?

HONEYBEE

BUZZING HIVE BUILDER

This insect can build nests almost anywhere. Put your hands together for the hardworking honeybee!

Antennae to sense smells

HONEYBEE STATS

HOME
Grasslands and forests all over the world

FOOD
Pollen, nectar, and honey

SIZE
About ½ inch (1 cm) long

A HONEYBEE IS SHORTER THAN A CRAYON.

SOCIABLE WEAVER

ENDURING DESIGNS

This bird's nests can last for more than 100 years. Make some noise for the super sociable weaver!

SOCIABLE WEAVER STATS

HOME

The Kalahari Desert of southern Africa

FOOD

Insects and seeds

SIZE

5½ inches (14 cm) tall

A SOCIABLE WEAVER IS SMALLER THAN AN AMERICAN ROBIN.

COLONY CONTEST

HONEYBEE VS SOCIABLE WEAVER

HONEYBEE

Honeybees build hives inside hollow trees. They also build hives between rocks. Bees raise their young in the hives.

HOW IT'S MADE

Worker bees make wax in their bodies. They use the wax to make cells. Bees connect the cells to build the hive.

THAT'S BUSY!

A colony can have up to 100,000 bees.

SOCIABLE WEAVER

Sociable weaver nests have hundreds of small rooms. The birds raise their babies here.

HOW IT'S MADE

Sociable weavers build nests with twigs and grass. The birds add on to the nests regularly.

THAT'S BUSY!

Up to 100 sociable weaver families can live in one nest.

GLOSSARY

algae—a plant or plantlike organism that lives in water.

attract—to cause someone or something to come near.

bower—a structure made from intertwined sticks and twigs.

dung—animal waste.

fungus—an organism, such as mold or mildew, that grows on rotting plants.

rodent—a mammal with large, sharp front teeth, such as a rat, mouse, or squirrel.

saliva—a liquid produced in the mouth.

savanna—a grassland with few trees.